NEW VANGUARD • 144

UNMANNED AERIAL VEHICLES

Robotic Air Warfare 1917–2007

STEVEN J ZALOGA ILLUSTRATED BY IAN PALMER

First published in Great Britain in 2008 by Osprey Publishing,
Midland House, West Way, Botley, Oxford, OX2 0PH, UK
443 Park Avenue South, New York, NY 10016, USA
E-mail: info@ospreypublishing.com

A CIP catalog record for this book is available from the British Library

ISBN: 978 1 84603 243 1

Page layout by: Melissa Orrom Swan, Oxford
Index by Alison Worthington
Typeset in Sabon and Myriad Pro
Originated by PDQ Digital Media Solutions
Printed in China through Worldprint

08 09 10 11 12 10 9 8 7 6 5 4 3 2 1

AUTHOR'S NOTE

The author would especially like to thank Dana Bell for providing historical photos as well as information on UAV camouflage colors.

EDITOR'S NOTE

For ease of comparison between types, imperial measurements are used almost exclusively throughout this book. The following data will help in converting the imperial measurements to metric:

1 mile = 1.6km
1lb = 0.45kg
1 yard = 0.9m
1ft = 0.3m
1in. = 2.54cm/25.4mm
1 gal = 4.5 liters
1 ton (US) = 0.9 tonnes

UAV TERMINOLOGY

The terminology for military drone aircraft has changed over the years. In the earliest days, they were referred to as "pilotless aircraft," but this term also referred to aircraft used as early cruise missiles. The most common term in the 1940s and 1950s was drone, or drone aircraft, since they were usually controlled from another aircraft. By the 1960s, the term RPV (remotely piloted vehicle) emerged since they could be controlled from the ground or air. The present term, UAV (unmanned aerial vehicle), appeared in the 1980s, although in recent years this has sometimes mutated into UAS (uninhabited aerial system). Armed drones were first called "assault drones" in World War II, and more recently, UCAV or UCAS (uninhabited combat air vehicle/system). In the civil aviation world, they are termed "ROA" (remotely operated aircraft) under US Federal Aviation Administration statutes.

CONTENTS

UNMANNED AERIAL VEHICLES
ROBOTIC AIR WARFARE 1917–2007

INTRODUCTION

On the night of November 3, 2002, in the remote Marib desert of Yemen, an automobile carrying a senior Al Qaeda leader suddenly exploded. The automobile had been destroyed by a Hellfire missile fired from a Central Intelligence Agency (CIA) RQ-1A Predator drone flying thousands of feet above, controlled from a clandestine site hundreds of miles away in Djibouti. The attack ushered in a new age of robotic air warfare.

The use of robotic aircraft long predates the recent wars in the Middle East and can be traced back to before World War II. The first attempts at pilotless aircraft took place in 1917 with early experiments in radio control and inertial guidance. The first practical drone aircraft were used as targets for training antiaircraft gunners. World War II saw the first attempts to use assault drones in combat, but this technology evolved mainly in the direction of guided missiles. By the 1950s, drones had been adapted to conduct aerial reconnaissance, which would become their principal mission for the next half century. The first large-scale use of UAVs (unmanned aerial vehicles) was during the Vietnam War where they flew thousands of spy missions too hazardous for manned reconnaissance aircraft. With the arrival of new technologies such as digital cameras, satellite navigation, and computer microprocessors, the capabilities of the robotic aircraft have increased enormously. Their variety has increased as well, from mini-UAVs the size of a model airplane to endurance UAVs with the wingspan of a modern jumbo jet. Indeed, the capabilities of UAVs have developed to such a point that many air forces today are wondering if robotic aircraft will replace piloted aircraft in the next generation of warplanes. In spite of their advances, robotic aircraft are still in their infancy. This book will examine the dawn of robotic air warfare.

THE EARLY DAYS

Remotely piloted aircraft first appeared during World War I, but the early efforts were stymied by the primitive guidance technology available. To pilot an aircraft remotely, some form of flight control system was needed as a substitute for the pilot. In these early days, UAV and guided missile development were intimately connected, since the challenges of flight guidance were essentially the same between both types of aerial vehicles. Guided missiles differ from UAVs in one crucial respect: UAVs are designed to return to base after their mission, while guided missiles explode when they impact their target.

OPPOSITE
Target drone technology nurtured early UAV development. The clearest example was the Vietnam-era "Lightning Bug" that stemmed from Teledyne-Ryan's successful BQM-34 Firebee. (Teledyne-Ryan)

Will future air combat be dominated by robotic warplanes? This illustration shows Boeing's X-45 UCAV (uninhabited combat air vehicle). (Boeing)

In 1909, the American inventor Elmer Sperry began designing gyroscopic devices to control the stability of aircraft in flight, these being ancestors to modern inertial navigation systems. The US Navy showed some interest in this concept to create an "aerial torpedo," a precursor of modern cruise missiles. To improve the accuracy of the aerial torpedo beyond the very limited capabilities of early gyros, radio control was developed by the Western Electric Company. These two technologies, inertial navigation and radio control, would form the core of remotely piloted aircraft development over the next 80 years. The first flight of the Curtiss-Sperry Aerial Torpedo took place in December 1917. The US Army Air Force sponsored a competitor, the Liberty Eagle Aerial Torpedo developed by Charles Kettering. In Britain, Sopwith, DeHavilland, and the Royal Aircraft Factory all attempted similar programs, but none successfully flew. By 1918, it was evident that the technology of the day was not adequate to create a viable guided weapon, and the programs petered out. However, both the US Navy and Royal Navy realized that remotely piloted aircraft could serve as realistic target drones for antiaircraft gunners. As a result, target drones became the principal form of pilotless aircraft for nearly a half century.

The development of robotic aircraft between the two world wars mainly concentrated on primitive guided missiles and target drones. The Royal Aircraft Establishment began test flights of its RAE 1921 Target Aircraft in 1922 and adapted it as an aerial torpedo as the Larynx (Long Range Gun with Lynx Engine). In 1920, the US Army began sponsoring a program called the Messenger to develop an inexpensive aircraft that could be used to deliver messages between headquarters in place of runners. As a futuristic spin-off from this program, the Army approached Sperry to develop a robotic version called the Messenger Aerial Torpedo (MAT), which could remotely fly from one headquarters to another. Tests were conducted starting in 1920, and improvements were gradually added including radio control to supplement the inertial navigation system. Although a remarkable technical achievement and a milestone in the history of aviation navigation technology, the MAT concept was far beyond the capabilities of the day, and funding ended in 1926.

The Royal Navy was one of the early proponents of target drones, first flying the radio-controlled Fairey Queen in 1933 and then acquiring over 400 DH 82B Queen Bee target drones from 1934 to 1943, based on the ubiquitous Tiger Moth biplane trainer. The use of target drones in the United States emerged from the hobby industry in the 1930s with Reginald Denny and his Radioplane Company. Denny used his experience in designing remote control model airplanes to design his Radioplane-1 (RP-1), but the US military showed no interest until 1939, on the eve of war. Some 15,374 of the various Radioplane drones from RP-4 to RP-18 were built during World War II. The improved RP-19/OQ-19 appeared in 1946, and over 48,000 were built from 1946 to 1984. In 1952, Radioplane was acquired by Northrop and went on to form the core of one of the most successful of today's UAV firms.

In Germany, Dr. Fritz Gosslau of the Argus Motor Works developed the FZG-43 (Flakzielgerat-43, antiaircraft target device-43) for training Luftwaffe flak crews. In October 1939, Argus proposed a more revolutionary scheme using a larger radio-controlled drone dubbed Fernfeuer (Deep Fire). It could carry a one-ton bomb load and would be controlled by a piloted version of the same aircraft. On delivering its bomb, the Fernfeuer would return to base. This was not an aerial torpedo/cruise missile like the Sperry design but rather an ancestor of today's UCAV (uninhabited combat air vehicle). Although the Luftwaffe ignored the Fernfeuer, this program laid the groundwork for the FZG-76, better known as the V-1 cruise missile.

THE FIRST ASSAULT DRONES

The US Navy had continued to experiment with larger radio-controlled aircraft as target drones in the late 1930s, and they were deployed with Utility Squadron Five (VJ-5) at Cape May, New Jersey, in March 1941 to train Navy antiaircraft gunners. The VJ-5 commander, Lt. Robert Jones, suggest using the drones as aerial rams to attack enemy fighters, eventually leading to the

The Kettering Bug was one of a number of attempts to develop pilotless aircraft at the time of World War I. The "aerial torpedo" was in fact an ancestor of the modern cruise missile rather than UAVs, but these initial attempts at inertial navigation and radio control were essential to the advance of pilotless flight. (USAF Museum)

The US Navy's assault drone program began with the TDN-1. It was fitted with a rudimentary cockpit for a pilot during ferry flights, but this was faired over when used in combat. Weapons, such as the two depth charges seen here, were carried underneath. (NARA)

Navy's Gorgon antiaircraft missile program. In the meantime, the Navy had been sponsoring the development of alternative flight control and navigation technologies including RCA's television camera and the Navy Research Lab's (NRL) radar guidance system. These offered the possibility of guiding the drone much more accurately than radio control, and in 1941 a program was started to develop "assault drones." These could be used either as guided missiles, impacting against an enemy target, or as UCAVs, dropping a weapon and then returning to base. The Navy ordered the production of the first TDN-1 assault drones by the Naval Aircraft Factory in March 1942. Since the drone was expendable, the Navy wanted a simpler and less expensive type, which was manufactured by Interstate Aviation as the TDR-1. The Navy's top secret Operation *Option* envisioned as many as 18 attack drone squadrons, 162 TBF Avenger control planes, and 1,000 assault drones. This ambitious scheme was whittled down considerably, and in March 1944, two Special Air Task Force (SATFOR) squadrons were dispatched to the Pacific Theater to demonstrate their capabilities. The TDR-1 were initially used as guided missiles at Bougainville in September 1944, flown into Japanese bunkers and gun positions. On October 19, 1944, they were used for the first time in a UCAV configuration, dropping bombs on targets on Ballale Island, south of Bougainville. They were not UCAV in the contemporary sense, since they lacked the capability to return safely to base. The Navy was not overly impressed with the results, although a similar concept was tested in the Korean War using radio-controlled Hellcat fighters as primitive missiles.

Through most of World War II, there was a very close connection between early guided missile development and various types of target drones and assault drones. Several countries developed radio-controlled bombers to crash into high-value targets: programs included the US Aphrodite program, the

A **INTERSTATE TDR-1 ASSAULT DRONE, US NAVY SPECIAL TASK AIR GROUP ONE, SOLOMON ISLANDS, AUTUMN 1944**

When deployed to the Pacific Theater of Operations in May 1944, the TDR-1 assault drones were still finished in their original scheme of dark sea blue over white. Aside from the usual national insignia, markings consisted of the occasional name on the nose and sometimes a tactical number painted on the tail. During combat operations, the forward cockpit windscreen was removed, and the cockpit was faired over with a plate, as seen here. After takeoff, the landing gear was dropped since the TDR-1 was on a one-way mission. The TBM-1C Avenger control aircraft was modified with a large receiver/transmitted antenna under the rear fuselage contained in a dome-shaped cover.

The first tactical UAV was the US Army's SD-1, later called the MQM-57 Falconer, a derivative of the prolific family of Radioplane target drones in use since World War II. It was fitted with a simple camera and flew back to base for recovery and film processing. (USAF Museum)

German Mistel, and the Italian Assalto Radioguidato. World War II also saw the development of the first dedicated guided missiles using inertial, TV, radio, and radio command technologies such as Germany's Fritz-X guided bomb, Hs-293 antiship missile, and the US Navy's Bat antiship missile.

COLD WAR ROBOT SPIES

Drones were used in increasing numbers for target training through the 1950s, but most pilotless aircraft development in these years was focused on the burgeoning new technology of guided missiles. In the mid-1950s, the US Army began to show interest in modifying target drones to carry small cameras to carry out battlefield reconnaissance. The first of these was the SD-1 (Surveillance Drone), based on the Radioplane RP-71 target drone. This carried either a KA-20A daylight camera that could take 95 photos or the KA-39A infrared night camera capable of taking ten photos. The SD-1 drone was launched using rocket-assisted takeoff (RATO) and was tracked by radar while its pilot controlled it via radio commands during its 30-minute flight. It flew back to base and was recovered by parachute. The drone and its associated equipment were designated as the AN/USD-1 and was the world's first successful surveillance UAV. It pioneered most of the critical technologies used for the past half century and is the precursor of today's reconnaissance UAVs. It was typically deployed in an aerial surveillance and target acquisition platoon within the division's aviation battalion with 12 drones. Some 1,445 were built, serving in the US Army from 1959 to 1966. The British Army also acquired this drone, where it was known as the Observer.

The USD-1 program was followed by a very ambitious program to field more sophisticated surveillance drones, but most were cancelled by the mid-1960s due to technical problems or excessive cost. One of the major shortcomings with these programs was the inconvenient and time-consuming methods needed to process the intelligence information. After the drone returned, its camera had to be unloaded and the film processed. This procedure might be acceptable for fixed strategic targets such as factories, but in the case of typical army objectives such as enemy tank formations or tactical missile launchers, the target would probably have moved between the time the photo was taken and the time that the film was processed and distributed. While television cameras might have solved this problem by providing a real-time solution, the early TV cameras were too cumbersome.

The US Air Force had a few surveillance drone programs at this time, including the Northrop B-67 Crossbow, but it was never fielded due to its high cost. In the early 1950s, the USAF also began examining drones to serve as decoys that could be launched into Soviet airspace in the vanguard of a bomber attack to draw away fighters and missiles. After examining the rocket-powered Convair GM-71 Buck Duck and the ground-launched Fairchild SM-73 Blue Goose, the first decoy drone actually to enter service was the McDonnell Douglas GAM-72 Quail in 1961. This was a small jet-powered drone launched from a B-52 bomber. About 600 were manufactured, and they served until 1972, by which time Soviet radar technology had improved to the point that they were no longer effective decoys. Its successor, called SCAD (Subsonic Cruise Armed Decoy), evolved into the AGM-86 ALCM (air-launched cruise missile) when it was realized that the decoy should carry a warhead as long as it was flying in hostile airspace. Decoy drones are a specialized category of air vehicle falling between UAVs and missiles, and development has continued on a modest scale ever since.

The shoot-down of a CIA U-2 spy plane over the Soviet Union in May 1960 by a Soviet S-75 (SA-2) missile dramatically changed the US interest in robotic spy aircraft to avoid the political embarrassment of a captured pilot. Two programs were started in response to this requirement, the Air Force's "special purpose aircraft" (SPA) program and the CIA's D-21 Tagboard supersonic drone program. The SPA program was another target drone spin-off, based on the widely used Ryan Firebee. Ryan had promoted the idea of a

The Teledyne-Ryan AQM-91A Compass Arrow was an evolutionary attempt to improve on the Firebee drone, by moving the engine air intake to the top of the fuselage to make it less visible to radar. In addition, the wings were extended to provide better endurance and high-altitude performance up to 80,000ft. It was intended for use over China but was never put into production due to improving international relations. (USAF Museum)

Drone aircraft can be employed as decoys to bluff enemy surface-to-air missiles, and the first type to be deployed was the McDonnell Douglas GAM-72 Quail, carried on B-52 strategic bombers in the 1960s. (Author)

reconnaissance version of the drone as early as 1959, but it took the U-2 shoot-down to interest the USAF. In summer 1960, Ryan began to build a specialized version of the Firebee drone to reduce its radar signature, increase its range, and improve its flight controls. A lack of funding delayed the program, but the loss of an Air Force U-2 to a SA-2 missile during the Cuban missile crisis in autumn 1962 reinforced the urgency of the program. After four Ryan 147A drones had been tested, the USAF proceeded with the improved Ryan 147B, which extended the wing for better range. This was only the start of a whole family of Firebee reconnaissance drones with specialized features and improvements. The Soviet SA-2 missile was viewed as such a serious threat that the Model 147D was specifically developed to help learn its secrets. It carried a "SAM sniffer" electronics intelligence (ELINT) package to pick up the distinctive electronic signals of the SA-2 guidance system. The Ryan 147D could be flown in harm's way to bait a missile site, then relay the electronic signals back to a waiting ERB-47 electronic warfare aircraft standing off at a safe distance. For security reasons, the Firebee drones were rechristened as "Lightning Bugs" before they made their operational debut.

LIGHTNING BUGS OVER VIETNAM

After the Republic of China Air Force (RoCAF) destroyed more U-2 spy planes over the People's Republic of China, the region became the first theater of operations for the Lightning Bugs. The first five missions were flown over China in early September 1964, and two were successful. Although more missions were flown over China in the ensuing months, most of the Lightning Bug missions were directed to a new battle zone when the Vietnam War began to heat up. The high-altitude Lightning Bugs were not especially useful during the monsoon season from November through March as most of the target areas were cloud covered. As a result, the new low-altitude Ryan 147J was developed specifically for missions over North Vietnam.

B

LOCKHEED MARTIN D-21B TAGBOARD, US CENTRAL INTELLIGENCE AGENCY, 1970

The D-21 Tagboard supersonic reconnaissance drone program was a joint US Air Force-CIA venture. Due to its clandestine nature, the aircraft was sparsely marked, in part to prevent obvious attribution if the aircraft crashed in China on one of its missions. In operational use, it was painted in overall black, possibly special radar-absorbent paint.

B LOCKHEED MARTIN D-21B TAGBOARD

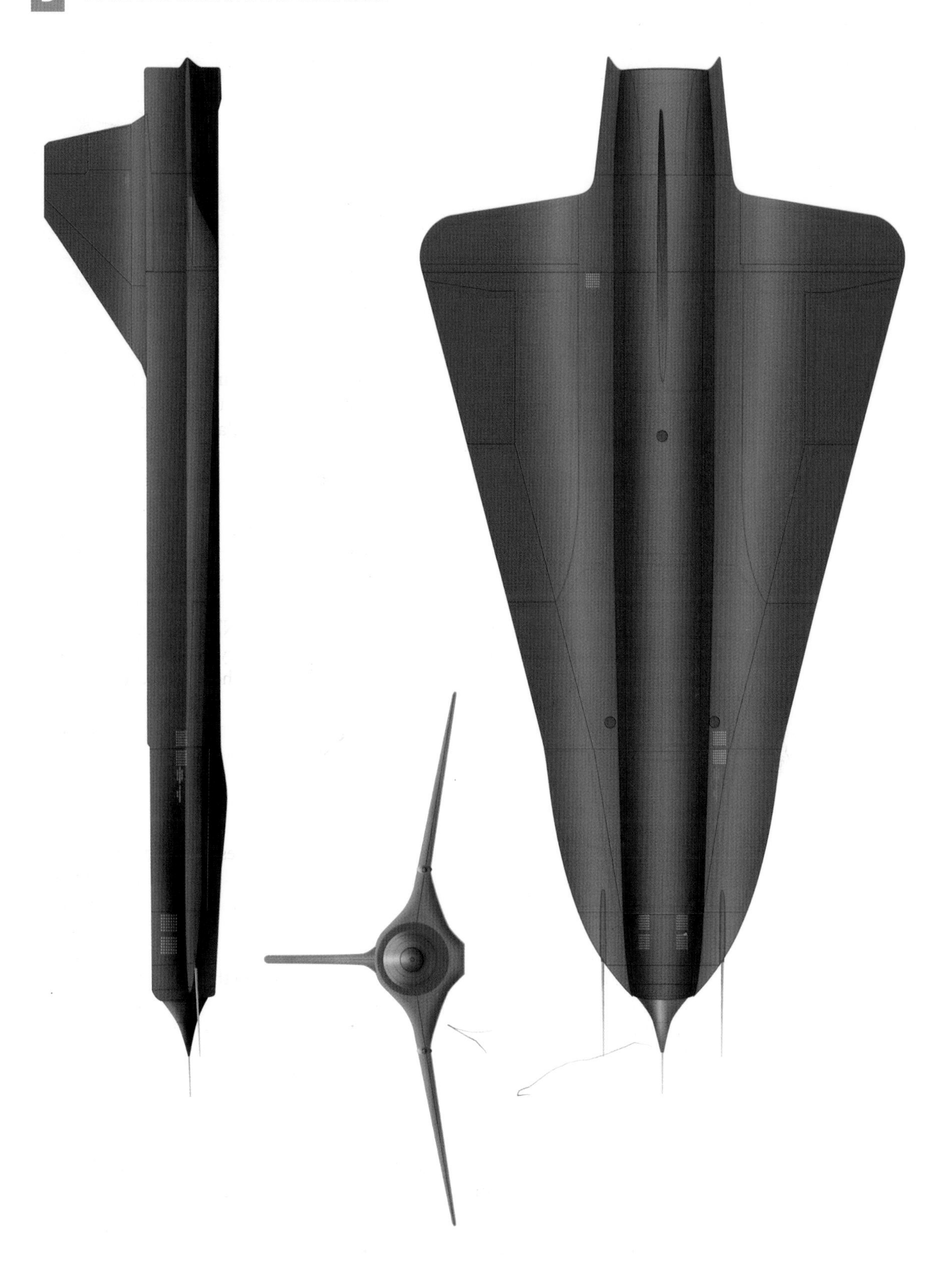

Alone, unarmed, unafraid. The first major combat use of UAVs came during the Vietnam War where they were extensively used to conduct reconnaissance missions too dangerous for manned reconnaissance aircraft. Remarkably, this Model 147 Lightning Bug survived nearly three dozen missions over Vietnam and was refurbished to serve as a target drone in 1988 off the coast of Bangor, Maine, to test the air defenses of the USS *Philippine Sea* AEGIS-class cruiser. (Teledyne-Ryan)

Lightning Bugs of the USAF 100th Strategic Reconnaissance Wing began missions over North Vietnam in late 1965 and had their first major success on February 13, 1966, when a Ryan 147E fitted with a new SAM sniffer finally managed to detect the SA-2 missile command link signal seconds before it was destroyed by the missile. One of the more intriguing techniques to retrieve the drones was to snatch them and their parachutes using a helicopter-based Mid-Air Retrieval System (MARS), fitted first to the CH-3E Little Jolly and later to the CH-53 helicopters. Some 23 versions of the Lightning Bug were developed at the time of the Vietnam War, including night reconnaissance versions with special strobe flashes to illuminate the target area, SIGINT (signals intelligence) versions, and electronic antiradar jamming versions. In the decade from 1964 to 1974, 1,016 Lightning Bugs flew a total of 3,435 sorties over China, North Vietnam, and North Korea. Of these, 544 Lightning Bugs were lost of which about one-third crashed due to mechanical problems while the rest were shot down by antiaircraft guns, jet fighters, or SA-2 missiles. The drones were credited with the loss of several North Vietnamese MiG fighters, either when the MiG crashed trying to intercept them or during missile engagements when the North Vietnamese missile missed the tiny drone and hit another fighter instead. One drone was credited with "ace" status as it was involved in the loss of five North Vietnamese fighters. North Vietnamese SA-2 crews claimed to have downed 130 Lightning Bugs, but this is probably an exaggeration; North Vietnamese MiG pilots claimed 11. The anticipated life expectancy of a Lightning Bug in combat over Vietnam was only 2.5 sorties, but the actual average was 7.3

The secret and futuristic Lockheed D-21B Tagboard was operationally flown several times in the 1970s to monitor developments at China's Lop Nor nuclear test range, but none of the missions succeeded due to the complexity of recovering the film capsule. (USAF Museum)

missions. The record was set by a Model 147S named "Tom Cat" that flew 68 sorties before being lost on September 25, 1974. The Vietnam War was the first large-scale use of drones in combat.

Another drone developed during this period was one of the most secret projects of its day – and one of the most technologically sophisticated. The Lockheed Skunk Works had completed design of the A-12 spy plane for the CIA as a supersonic replacement for the U-2. It was becoming evident that advances in Soviet air defense missiles would make it increasingly dangerous for piloted aircraft to fly into heavily defended regions of the Soviet Union and China. As a result, in October 1962, the CIA funded Lockheed to begin work on the Tagboard reconnaissance drone based on the technology of the A-12. The modified version of the A-12 carrying and launching the drone was called M-21 and the drone D-21; "M" and "D" for "Mother" and "Daughter." The D-21 was carried on a pylon over the spine of the A-12, and two were modified into M-21 mother ships. The D-21 was capable of speeds in excess of Mach 3 and had a range of 3,450 miles. Unlike other reconnaissance drones of the period, the D-21 was not recoverable but rather it shed a camera pod that was recovered at sea.

The first test flight was conducted on March 5, 1966. The third test flight on July 30, 1966, led to a collision between the M-21 and drone due to separation problems, resulting in the loss of both craft and one of the crewmen. The danger of this launch method led to substitution of the B-52H bomber as the mother ship, as the modified D-21B drone was launched from a more conventional pylon under the wing. After several failed tests, the first successful flight from a B-52H took place on June 16, 1968. The first "Senior Bowl" mission was conducted on November 9, 1969, with the objective being the Chinese nuclear test facility at Lop Nor. Although the D-21B completed its mission over Lop Nor, it failed to make the turn back to the recovery site and crashed in Mongolia. A second mission, on December 16, 1970, was successful until the final phase in the recovery area when the drone failed to eject the camera pod. The third mission, on March 4, 1971, was even more frustrating. The drone completed the mission and successfully deployed the film pod; unfortunately the destroyer sent to recover the pod ran over it in the recovery process and the pod sank. The fourth and final mission, on March 20, 1971, failed during the return trip to the recovery site when apparently it was shot down. The program was cancelled for a variety of reasons including the obvious technical problems, US rapprochement with China, and the growing capabilities of US spy satellites.

Mother and daughter: the CIA's M-21 mother ship and D-21 Tagboard drone on one of its early test flights. Separation between the mother ship and drone proved so dangerous that the M-21 concept was abandoned in favor of using B-52 bombers. (US DoD)

The DC-130 served as the mother ship for Firebee drones during the Vietnam War. This DC-130 is in use by the US Air Force's 6514th Test Squadron, hosting AQM-34 Firebee target drones to test air defense systems like those on *Aegis* cruiser USS *Chosin* seen below in May 1991. (US Navy)

SUB-KILLER DRONES

In the 1950s, the US Navy acquired the first helicopter attack drone, the QH-50 DASH (drone antisubmarine helicopter). The QH-50 was based on an earlier one-man helicopter design by the Gyrodyne Company that used two counter-rotating propellers for vertical lift. The US Navy began considering the possibility of using such a helicopter to carry torpedoes, since modern destroyer sonars were capable of detecting enemy submarines to ranges over 20 miles, which was beyond the reach of their weapons. The idea was that the destroyer could detect the submarine and then launch the QH-50 drone to deliver antisubmarine torpedoes at distances up to 20 miles from the ship. The QH-50 made its first flight on August 12, 1960, and it was first deployed on destroyers in January 1963. Through 1969, some 810 QH-50 were manufactured, and they served aboard 240 US Navy destroyers. The other major user was the Japanese Naval Self-Defense Force, which acquired 24 DASH drones. In US Navy service, the QH-50 proved to be a versatile device for missions other than its primary antisubmarine role, such as surveillance, gunfire correction, smokescreen laying, and transport of cargo. During the Vietnam War, some QH-50 were fitted with video cameras and used to conduct artillery-spotting missions. More advanced night vision sensors, such as the Nite Panther and Nite Gazelle, were also fitted, permitting the deployment of QH-50 night assault drones armed with a minigun, grenade launcher, submunitions container, bombs, or other weapons.

By summer 1970, some 441 QH-50 drones had been lost mainly due to peacetime accidents; only 5 percent were combat losses in Vietnam. The mean time between loss averaged only 145–185 hours of flight time. The main problem was the unreliability of the electronics. The Navy drones were retired in 1970–71, and they were replaced with the LAMPS helicopter. Surviving QH-50 drones remained in

use with the US Army for many years for towing aerial targets. Both Germany and Israel acquired a few for further exploration of the naval UAVs. It spite of its short service life, the QH-50 DASH was far ahead of its time and was the pioneer of naval and helicopter UAVs.

Introduced in 1962, the first Soviet tactical drone was the TBR-1 system that used the Lavochkin La-17R drone launched with rocket assistance from the SATR-1 rail launcher based on an antiaircraft carriage. The air vehicle was derived from a common Soviet target drone.

RED DRONES

As in case of the United States, the vulnerability of reconnaissance aircraft to new air defense missiles prompted the Soviet Air Force in the late 1950s to develop reconnaissance drones based on the widely used Lavochkin La-17 target drone. The TBR-1 (Takticheskiy bespilotny razvedchik, or unmanned tactical scout) system entered service in 1962, based on the widely used Lavochkin La-17 target drone. Four squadrons were formed: two in Ukraine, one in Belorussia, and one in Latvia. The system remained in service until the early 1980s. The La-17R drone could be flown either autonomously using an onboard navigation system or by remote control via radar tracking and a radio uplink. The system used a conventional film camera, and its operational radius was up to 125 miles when operating at an altitude of 23,000ft; less when operating at lower altitudes.

A far more ambitious design appeared in the early 1960s from the Tupolev Design Bureau. Tupolev had developed a ground-launched strategic cruise missile designated Izdeliye 121 (Item 121), a Soviet equivalent of the USAF SM-62 Snark missile. The program was killed in early 1960 as ballistic missiles promised to be more practical means for delivering strategic nuclear weapons. However, Tupolev was convinced of the basic merits of the design and promoted the idea of turning it into a long-range reconnaissance system, codenamed Tu-123 Yastreb (Hawk). The airframe was enormous and could house an array of cameras as well as an electronic intelligence system. The

OPPOSITE
The QH-50 DASH was the pioneer helicopter UAV. Besides being used by the US Navy for antisubmarine warfare missions, it was used to demonstrate UAV attack capabilities under the Nite Gazelle program, fitted with day/night cameras and various armament options such as the XM129 grenade launcher. (US DoD)

The most enigmatic of the Soviet Cold War drones was the massive Tupolev Tu-123, part of the DBR-1 Yastreb system. There have been rumors that this supersonic drone was used in clandestine missions over NATO countries in the 1970s and 1980s. (Author)

flight system was autonomous, with the flight path programmed into the flight control system prior to launch. The drone was supersonic, with a cruising speed of 1,675mph (Mach 2) and a typical operating altitude of over 65,000ft. After completing the mission, the drone returned to a recovery area where a ground signal activated the flight termination sequence. The drone dumped remaining fuel, did a zoom climb to bleed off speed, and then ejected the nose section containing the camera and ELINT sensors. The main portion of the airframe crashed, much like the contemporary US D-21 Tagboard drone.

In view of US overflights of Russia with the U-2 spy plane, the Kremlin proved enthusiastic about the idea, and the program was authorized in August 1960. Since the basic airframe had already been developed, flight trials soon followed, starting in 1961. The DBR-1 (Dalniy bespilotniy razvedchik, or long-range unmanned scout) was accepted for service in May 1964 and was in production from 1964 to 1972 with a total of 52 drones being built. They were used to equip three squadrons stationed in Ukraine, Belarus, and Latvia. Each squadron had six launchers and typically had a dozen drones on hand. The system was expensive to operate since each mission involved the loss of an airframe. A fully recoverable version called DBR-2, with the Tu-139 drone, was studied in the mid-1960s but did not proceed to the production phase. The DBR-1 squadrons were retired in the early 1980s. Although operated by the Soviet Air Force, they were subordinated to the GRU military intelligence service, and so there is some mystery regarding their actual missions. Russian defectors claimed that they were used to overfly Spain, Britain, the coast of France, and China.

By the late 1960s, it was obvious that the TBR-1 was obsolete and too vulnerable, while the DBR-1 was so expensive that it was sequestered for only the most important missions. Tupolev proposed the development of a pair of UAVs, the short-range Tu-143 and the long-range Tu-141. The tactical

DBR-1 YASTREB, SOVIET AIR FORCE 4TH IND. GUARDS UAV SQUADRON, VLADIMIR-VOLYNSKIY AIR BASE, 1973

The DRB-1 Yastreb system is shown here during a launch of the Tupolev Tu-123 reconnaissance drone from the standard SURD-1 (ST-30) semitrailer launcher by the 4-ya OGEBSR (4-ya Otdelnaya gvardeyskaya eskadrilya bespilotnikh samoletov-razvedchikov). The Tu-123 was generally left in a bare metal finish, although with red trim markings and the standard national insignia of the red star with thin red trim. Other markings were usually limited to a production serial number on the fuselage nose, often repeated on the side aft the wing.

The most widely used Soviet spy drone of the Cold War was the VR-3 Reys based on the Tu-143. This shows the improved VR-3D Reys-D with the Tu-243 drone and the SPU-243 launch vehicle behind. (Author)

VR-3 Reys (Flight) system was authorized in August 1968 and was based on the Tu-143. This drone was a subsonic canard design with the jet engine intake above the fuselage to reduce its radar signature. The Tu-143 had a maximum speed of 575mph and an endurance of 13 minutes and was intended to cover targets to an operational depth of 35–45 miles behind enemy lines. Two different sensor packages were developed, one using film cameras and the other using a video camera. The drone was fired from a SPU-143 launcher truck using RATO; on completing its mission it flew back to a recovery area where it deployed a parachute. State testing was completed in 1976, and the Tu-143 was in production through 1989 with about 950 drones being manufactured. It was replaced by a substantially modernized version, the Tu-243, as part of the upgraded VR-3D (Reys-D) system, which entered service in 1982.

The VR-3 systems were deployed under Soviet Air Force control for operational reconnaissance at front level. A drone reconnaissance squadron had four SPU-143 launcher vehicles and 12 Tu-143 drones, with a capability of launching about 20 missions per day. The system is still widely used by the Russian Air Force and by other air forces of the former Soviet Union. The VR-3 was exported to several of the former Warsaw Pact countries in the early 1980s. Czechoslovakia received two squadrons in 1984, and it was also exported to Romania. The VR-3 was supplied to Syria in 1984 and was apparently used over Israel and Lebanon.

The operational VR-2 Strizh (Swift) was based on the larger Tu-141 drone. Although the layout of the Tu-141 was very similar to the Tu-143, it was considerably larger and offered supersonic cruise with a speed of up to 810mph (Mach 1.1) and an operating range of 630 miles. The first flight was conducted in December 1974, and initial production began in 1979. The Tu-141 was launched from a semitrailer using RATO assist and was recovered much like the Tu-143 with a parachute. It used an autonomous flight control system, and was fitted with a panoramic camera and an oblique film camera. A total of only 152 were manufactured, and they were deployed with Soviet Air Force squadrons in the western republics facing NATO.

The VR-2 Strizh was based on the Tupolev Tu-141 drone and was launched using rocket assist from this SPU-141 semitrailer launcher. (Author)

A third Tupolev drone program of the early 1970s was the most mysterious. Codenamed *Voron* (Raven), it was an attempt to reverse engineer the Lockheed D-21 drone that had crashed in Mongolia during its first operational mission on November 9, 1969. The program was officially initiated in March 1971, and the Tu-95K bomber was its expected mother ship. It would appear that the program never progressed to the stage of flight tests, although details of the program remain secret.

ISRAELI INNOVATIONS

The Israeli Air Force (IAF) had watched US tactics in Vietnam with interest. Although Egypt had deployed the SA-2 missile in the Sinai at the time of the 1967 war, it had seen little use. Nevertheless, the IAF expected that surface-to-air missiles (SAM) would be a significant threat in the future. The role of the Firebee drones in Vietnam led the IAF to purchase one dozen Ryan 124I drones in 1971, locally called Mabat (Observation). Israel also acquired

Israeli Aircraft Division's Malat division was one of the pioneers of the new generation of tactical UAVs that could provide real-time intelligence using commercial video cameras. The Searcher was a further evolution of the Scout, which was used in the 1982 Lebanon war, and proved to be one of the most widely exported UAVs of the 1990s. (Author)

a reconnaissance version of the BQM-74 Chukar drone, and these two types were used to form the 200th Drone Squadron. They were used in a reconnaissance role in the 1973 Yom Kippur War. The difficulties posed by Egyptian and Syrian SAM sites in 1973 led to more interest in SEAD technology (suppression of enemy air defense), and the IAF began testing a variety of techniques to overcome these defenses. One was an unpowered decoy drone called UAV-A, which could be launched from fighters to fool SAM radars into thinking they were being approached by a massed formation of strike aircraft. In the late 1970s, Israel purchased the Brunswick Model 290P Propelled Decoy from the United States, which was then manufactured under license in Israel as the Samson.

At the same time, two Israeli firms began promoting the idea of using mini-RPVs (remotely piloted vehicles) the size of a large model airplane with inexpensive new lightweight video cameras to provide real-time surveillance. The Israeli Aircraft Industries (IAI) Scout and the Tadiran Mastiff both offered a method to monitor the battlefield using flexible, low-cost air vehicles that could be operated easily from near the frontlines. This concept was first demonstrated in 1981 when the South African Army used the IAI Scout during Operation *Protea* in Angola. Both the Scout and Mastiff were in small-scale use at the time of the 1982 war in Lebanon. One of the main tactical challenges to IAF operations in Lebanon was the threat posed by a heavy concentration of Syrian SAM batteries in the Bekáa Valley. These were precisely located by using reconnaissance UAVs, although at least three drones were lost in the process. At the start of the 1982 Lebanon war, the IAF concentrated on neutralizing the Syrian SAMs. Decoy UAVs such as the Samson were launched toward the SAM sites, and once the Syrian radars were activated, they were struck both by ground-launched antiradar missiles and by air-launched missiles, all the while being observed by reconnaissance UAVs. These novel UAV tactics made short work of the SAM defenses and gave the IAF a free hand to support Israeli ground operations. The 1982 air operations rejuvenated interest in decoy drones and also popularized small

The mainstay of NATO UAV units through the Cold War was the Canadair CL-89 Midge and its descendent, the CL-289. The CL-289 Piver was extensively used in peacekeeping operations in the Balkans from 1996 with both French and German units. (MBDA)

One of the most dangerous moments for small tactical UAVs is the recovery phase. Some UAVs deploy a parachute and gently float to earth. Others such as the US Navy's Pioneer were recovered aboard US battleships by flying into a net; in this case aboard the USS *Wisconsin* in November 1990 during Operation *Desert Shield*. (US Navy)

UAVs with real-time video cameras. These offered a cheap entry into aerial reconnaissance and started a worldwide trend. While Israeli UAV exports in the 1980s and 1990s helped to push along the boom in UAVs, the simplicity of the basic technology encouraged novice aerospace countries to enter the field, with local efforts in Pakistan, India, Singapore, Iraq, Iran, and many other countries.

EUROPEAN UAVS

In spite of Britain's early role in the development of remotely piloted aircraft, sharp defense cutbacks in the 1950s led to relatively little work in this field. Most countries had acquired target drones, but few reconnaissance drones. The most significant European UAV from the 1960s to the 1990s had Canadian origins. In the 1950s, Canadair developed target drones used for missile testing. In the late 1950s, the Canadian Army was looking for a means to locate targets deep in the enemy rear to be attacked by the new Honest John artillery missiles. The British Army had a similar requirement, so in June 1963, both countries agreed to jointly fund Canadair to develop an artillery reconnaissance drone, called the CL-89 Midge. This was a modestly sized, jet-powered drone that carried an infrared line scanner to collect photographs

The ambitious MQM-105 Aquila was far ahead of its time, able to provide real-time intelligence day or night and to designate targets with a laser beam. But the project was cancelled in the early 1980s, setting back the US Army tactical UAV program by almost two decades. (Lockheed-Martin)

The RQ-2 Pioneer was deployed on US battleships during the 1991 war with Iraq, serving as spotter aircraft for its massive 16-in. guns, in this case being operated by Fleet Composite Squadron 6 (VC-6) on the fantail of USS *Wisconsin* during Operation *Desert Shield* in November 1990. (US Navy)

day or night. The Midge flew a preprogrammed course and returned to base by means of a radio beacon, finally deploying a parachute for recovery. The first test flight was conducted in March 1964, but development was delayed as the British Army wanted to make certain that the design was as durable and "soldier-proof" as possible to avoid the numerous problems plaguing other UAVs of this time period. Germany, Italy, and France later joined the program, and the first system was delivered to the Bundeswehr in 1972 under the NATO designation AN/USD-501. The life expectancy of the drones was about ten flights, though some survived as many as 40. A total of 500 drones were manufactured by Canadair through 1983. After the German CL-89s were retired, they were transferred to the Turkish Army.

Although the Midge proved to be a durable workhorse, the Bundeswehr felt that the range was inadequate for some missions. In 1976, Germany agreed to provide the bulk of the funding to develop an extended-range version, called the CL-289. The first test flight was conducted in March 1980. The CL-289 offered longer range and a more sophisticated guidance system. Its Corsaire infrared line camera could send back data to the ground control station (GCS) in near-real-time via a datalink, but the Zeiss daylight film camera required recovery and processing as in the earlier CL-89. A multinational production effort was organized among Canadair, SAT (Société Anonyme de Télécommunication) in France, and Dornier in Germany. A production contract was signed in 1987, the largest defense export in Canadian history at nearly one-half billion dollars. The CL-289 was also known by its NATO designation, AN/USD-502. In the French case, the CL-289 was part of the larger Piver (Programmation et Interpretation des Vols d'Engins de Reconnaissance), which included a related reconnaissance pod developed by Aerospatiale, carried on the Mirage F1CR aircraft. Deliveries started in 1990, and France purchased 55 air vehicles with two ground station systems, while Germany procured 188 air vehicles and 11 ground stations sets. It was widely used in peacekeeping operations starting in Bosnia in 1996 with the French 7th Artillery Regiment of the NATO IFOR (Implementation Force) and elsewhere in the Balkans through 2005, flying over 1,500 sorties. The CL-289 was significantly upgraded in 2003 with new Global Positioning System (GPS) guidance and digital features, and it was renamed as the AOLOS-289.

The Boeing YQM-94A Compass Cope B was a 1973 attempt to develop an efficient endurance airframe that could enable a reconnaissance drone to fly long distances one day or more. Until the advent of compact satellite communication systems, such endurance UAVs surpassed even the most state-of-the-art equipment used for normal operations. (USAF Museum)

TACTICAL UAVS: THE NEXT GENERATION

The CL-289 was not widely adopted outside Germany and France as many other armies wanted a system that provided real-time daylight imagery. The older generation systems such as the Firebee and CL-289 carried film cameras that were processed after the air vehicle returned to base, so that the data was often hours old before it reached commanders in the field. This was acceptable for fixed targets but useless in mobile warfare where many of the targets, such as tanks, artillery, and missile launchers, were mobile. Britain had planned to develop its own next-generation UAV, the Westland Supervisor, but this was cancelled before reaching production. In 1982, a new program, the Marconi Avionics Phoenix, was initiated, aimed at designing a more advanced system than the CL-289, with real-time capability. This proved to be a deeply troubled program, and even after it was finally fielded in 1998, Phoenix continued to suffer from a number of operational problems.

Britain was not the only country to have serious difficulties with the new generation of tactical UAVs in the 1980s. Lockheed proposed a new US Army UAV called Aquila in the early 1970s, and full-scale development started in 1979. The Aquila was part of a broader US Army effort to conduct "deep battle" using advanced new technology. Besides a real-time camera, the Aquila was also fitted with a laser designator, so it could hunt out high-value targets and then illuminate them with a laser beam, at which point the target could be attacked by the new laser-guided Copperhead 155mm artillery projectile. In spite of its enormous promise, the MQM-105 Aquila program was cancelled in 1987 after having run over budget. The program attempted to cram too many sensors into too small an airframe, and it was simply too far ahead of its time.

The US Navy and US Marine Corps did not want to wait for the futuristic Aquila, and when the Israelis demonstrated tactical UAVs in action in Lebanon in 1982, Secretary of the Navy John Lehman pushed the service into acquiring new UAVs off the shelf. The US firm AAI teamed with IAI on the Pioneer, which was based on existing Israeli designs, and it was selected over competitors in 1985. The AAI/IAI team

A critical technological innovation in endurance UAVs was the advent of systems to permit communication between the ground control station and the UAV via space communication satellites. Here is a Predator with the nose removed, exposing its satellite uplink antenna. (USAF)

would prove to be a durable partnership in American tactical UAV programs over the next two decades. The Pioneer could be launched either from ship or from land, using RATO for takeoff and parachutes or nets for recovery. The Pioneer was first deployed aboard the battleship USS *Iowa* in November 1987 while conducting patrols in the Strait of Hormuz in the Persian Gulf. The Pioneers were used to monitor Iranian Silkworm antiship missile sites and coastal traffic. The US Army began to explore the Pioneer as an alternative for the failed Aquila program. In spite of the success of the Pioneer, the US Department of Defense (DoD) halted any further purchases in 1988 in favor of developing a new common family of UAVs. This proved to be a major bungle as the Navy and Army had fundamentally different requirements for tactical UAVs, and a design that satisfied one service was unacceptable for the other. This mistake led to a decade of bureaucratic misadventures with neither the Army nor the Navy receiving a new tactical UAV. Instead, the Marine Corps Pioneers continued to toil away, remaining in service for more than two decades and still flying at the time this book was written.

OPERATION *DESERT STORM*

The war with Iraq in February 1991 saw the most extensive use of US UAVs since the Vietnam War and also saw their most varied use, with many different types in service. The Pioneer was deployed with the US Navy, US Marine Corps, and US Army during Operations *Desert Shield* and *Desert Storm*. The Navy operated the Pioneer from the battleships USS *Wisconsin* and USS *Missouri* for artillery adjustment and reconnaissance. Other Navy tasks for the Pioneer included mine reconnaissance and surveillance of Iraqi

THE FIREBEE DRONE IN COMBAT, 1969–2003

1: AQM-34L "Lightning Bug" (Ryan Model 147SC), USAF 99th Strategic Reconnaissance Squadron, 100th SR Wing, Vietnam, 1974

"Tom Cat" set the record for Lightning Bug sorties during the Vietnam War, conducting 68 missions before failing to return on September 25, 1974. Like most of the daytime, low-altitude Lightning Bugs, Tom Cat was finished in an overall scheme of light gull grey. Some dielectric panels were finished in a radio-transparent black or white paint such as the nose and tail antennas. The sortie markings depict a drone descending on a parachute. Some drones occasionally wore the "SAC Sash," a diagonal blue band with the Strategic Air Command insignia on a field of white stars. Tiger markings were always popular on the Lightning Bugs.

2: Ryan Model 124I Mabat, Israeli Air Force Squadron 200, Yom Kippur War, October 1973

The IAF Ryan 124I drones became operational with Squadron 200 at Palmahim Air Base in summer 1973. They were delivered in the usual light gull gray color with some dielectric panels in gray or black radio-transparent paint. Most of these drones were marked simply with a two-digit tactical number, but as seen here, some received more elaborate markings. The reason for the communist hammer-and-sickle marking on the fuselage is unknown, perhaps for deception if the drone was lost over Egyptian territory.

3: Northrop-Grumman BQM-34-53 Firebee, US Forces, Operation *Iraqi Freedom*, March 2003

Northrop-Grumman was given a quick-reaction contract in February 2003 to provide five BQM-34-53 target drones, modified to create chaff corridors during the US air attacks in Iraq. The modifications included incorporation of the larger BQM-34L wings for better lift and an improved navigation system with GPS satellite navigation for autonomous flight control. A set of special pods were fitted to the end of the wings, and a countermeasures dispenser was fitted in the nose. Three of these drones operated from US Navy DC-130A control aircraft while two were launched from the ground. They were painted in gunship gray, although as usual, some of the dielectric panels were in other colors, usually black.

D THE FIREBEE DRONE IN COMBAT

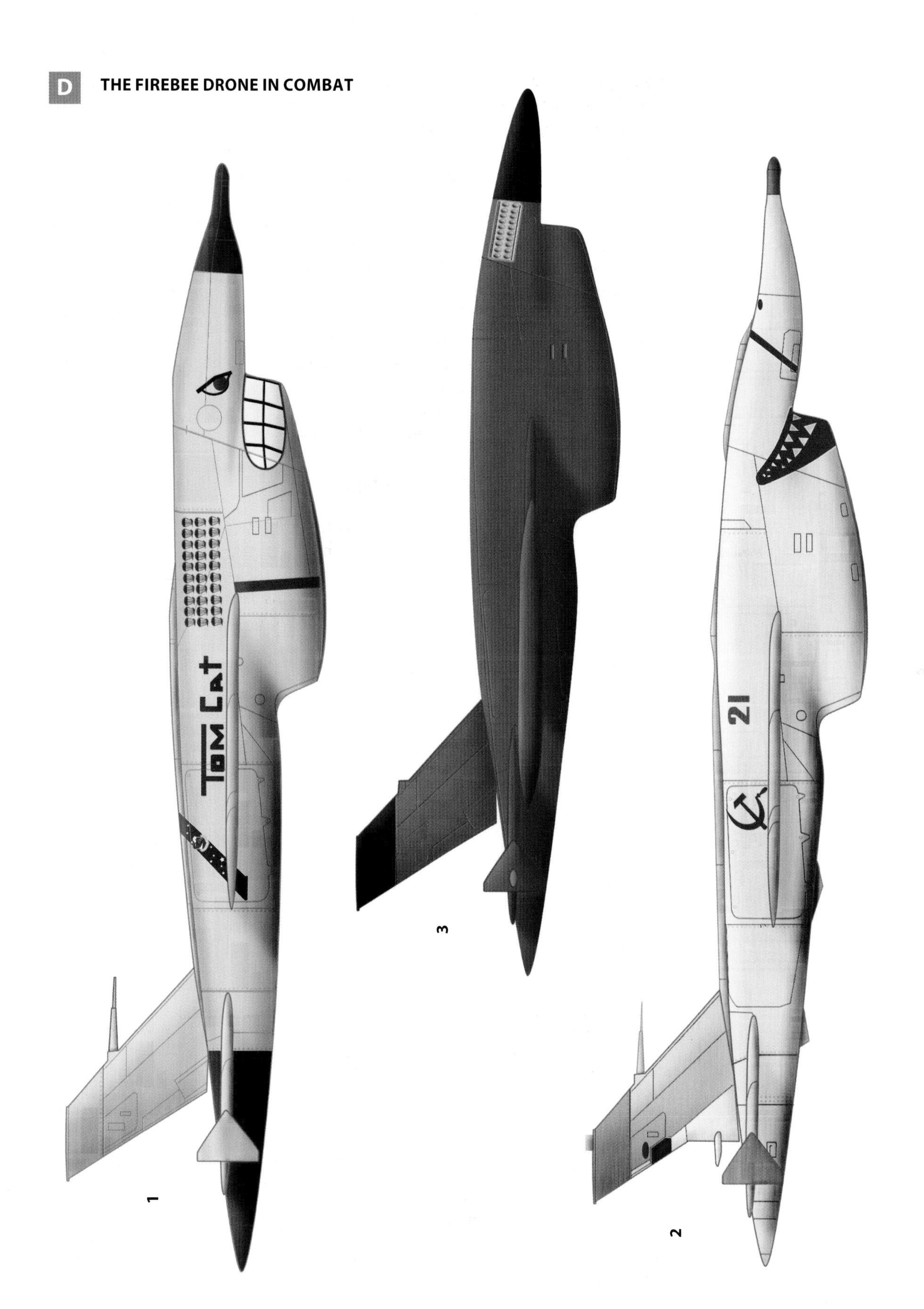

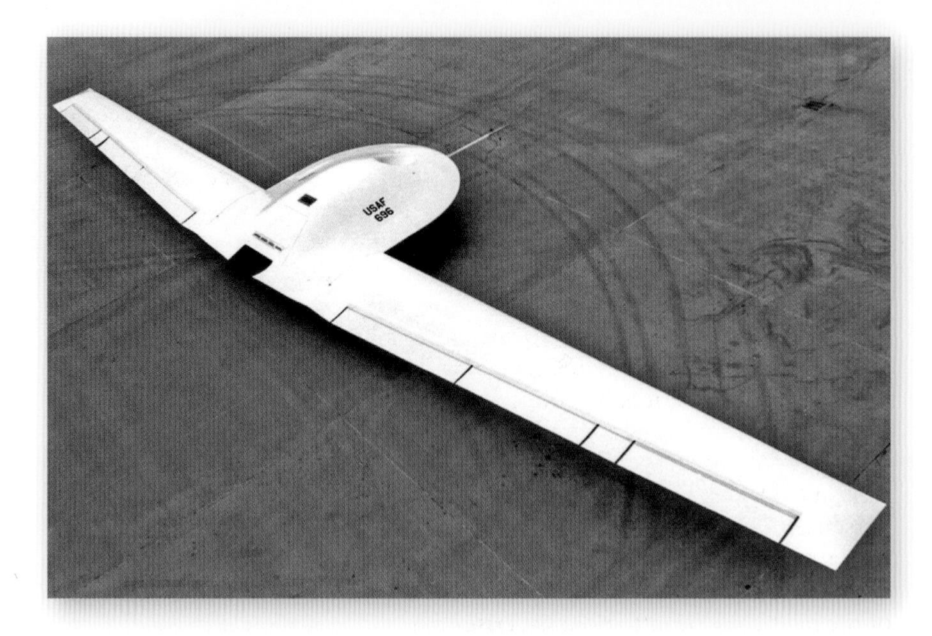

The most exotic of the three endurance UAVs developed for the US Air Force was the Lockheed Skunk Work's RQ-3 Tier 3- Dark Star, a stealthy flying wing design. (USAF Museum)

vessels to determine whether they were laying mines. The Iraqi Army quickly learned to fear the sounds of the little drones, as their presence was often followed by devastating battleship salvoes. In one of the more curious incidents of the war, an Iraqi unit surrendered to a Pioneer drone rather than endure another bombardment. Three Pioneer systems were deployed with the US Marine Corps with their 1st, 2nd, and 3rd RPV Companies. The 2nd RPV Company operated from the amphibious warfare ship USS *Guam*, while the other units operated from the ground. Gen. Alfred Grey, commandant of the US Marine Corps, stated that the Pioneer was "extraordinarily successful." The Pioneers flew a total of 523 sorties for 1,599 flight hours including 134 Navy and 346 Marine sorties. A single Pioneer system was employed by the UAV Task Force attached to the US Army 7th Corps. Among other tasks, the US Army used the Pioneers for route reconnaissance for Apache attack helicopters. A total of 40 Army Pioneer air vehicles were used, of which seven were lost and 20 damaged. Of the seven losses, two were to antiaircraft fire and five to accidents or hardware failure.

Although the US Air Force had extensive UAV experience in Vietnam, by the 1980s, this experience had largely been forgotten. Based on the Israeli experience in the 1982 war, there was still some interest in decoy drones to distract the Iraqi air defense radar net. During the opening phase of the Persian Gulf air campaign, the USAF's 4468th Tactical Reconnaissance Group fired 38 BQM-74C Chukar target drones into Iraq from ground sites in Saudi Arabia. The drones encouraged the Iraqis to turn on their air defense radars, which allowed trailing F-4G Wild Weasel aircraft, equipped with AGM-88 HARM missiles, to knock out significant portions of the Iraqi defensive SAM network. The US Air Force and Navy also used the smaller ADM-141 Tactical Air-Launched Decoys (TALDs) that were dropped from aircraft to create the appearance of false attacks.

The British Army used the dated Canadair CL-89 Midge with the artillery's 94 Locating Regiment. Since the CL-289 Piver was not yet ready, the French Army acquired a small number of off-the-shelf Altec MART Mk. II UAVs for use in Operation *Daguet*, the French portion of the Iraqi campaign.

The use of tactical UAVs in the Iraq campaign highlighted their value for surveillance and triggered considerable interest in most NATO armies. By the early 1990s, advances in commercial off-the-shelf technology considerably enhanced the capabilities of tactical UAVs. Miniaturized video cameras improved the collection capabilities of the UAVs and enhanced the quality of the video image. New advances in computer processor technology substantially improved UAV flight controls, permitting very sophisticated flight control systems to be built into very small UAVs. The new GPS navigation systems began to offer the promise of increased navigation accuracy.

These technological advances and the combat experiences in Iraq led to a marked upsurge in UAV deployments in the 1990s. In addition, tactical UAVs proved to be ideally suited to the first decade of the "New World Disorder" that followed the end of the Cold War. NATO armies were called on to conduct peacekeeping missions in the former Yugoslavia. UAVs were well suited to this mission since they could be used unobtrusively to observe ceasefires and to monitor the various militia forces. In addition, robotic aircraft posed no political embarrassment if shot down. In contrast, the loss of piloted reconnaissance aircraft over the Balkans with the ensuing capture of the pilot could lead to political crises that European governments wanted to avoid.

ENDURANCE UAVS

Two types of flight control systems were used by UAVs up through the 1980s: autonomous flight control and command guidance. Autonomous guidance was the earliest form of flight control and required that the air vehicle's flight control system be preprogrammed with a fixed course prior to launch. This was acceptable for early UAVs, which used film cameras, but it became less attractive once UAVs began carrying more capable sensors. Command

The enormous wingspan of the RQ-4A Global Hawk is evident in this overhead view of the second US Navy Global Hawk, en route to Edwards Air Force Base, California, on June 7, 2005. (USAF)

Elbit's Skylark IV was used by the Israeli Defense Force for patrols over Gaza and Lebanon during the turmoil in 2006. This design uses an electric engine and so is virtually silent in flight. (Elbit)

guidance involves a radio link between the pilot in the GCS and the air vehicle. Some of the early command-guided UAVs were tracked by radar, and the radar data allowed the pilot in the control station to make course corrections via the radio link. With the advent of televisions on UAVs, such as on some of the Lightning Bugs used in Vietnam, the UAV could send back an image to the pilot on the mother ship, and the pilot could then steer the UAV. By the 1980s, command guidance gradually became the preferred guidance method, although many UAVs have hybrid systems that use autonomous flight controls for a portion of the flight, combined with pilot commands during the intelligence collection phase.

The limitation of command guidance through the 1980s was the radio link. Radio signals could be disrupted by atmospheric conditions, equipment failures, or deliberate enemy jamming. In such an event, the loss of the radio link could lead to the UAV crashing or simply flying off, out of control. The radio link also placed a distinct limit on the range of the UAV. The USAF got around this problem to some extent in Vietnam by using a DC-130 mother ship that had a powerful transmitter. Ground-launched UAVs were far more limited than air-launched types since radio links were easily interrupted by geographic features such as mountains and had an inherent limit because UAVs would disappear over the horizon after about 40 miles Two technologies emerged in the 1990s to help circumvent this problem. Satellite uplinks could offer beyond-the-horizon capabilities even for ground-launched UAVs. The GCS would transmit its control signal up to a communications satellite in space, which would then be amplified and broadcast back down to the UAV via a special antenna in its nose. GPS satellite navigation also assisted in the problems caused when the command link was disrupted because the UAV

E

RQ-1 PREDATOR ENDURANCE UAV

1: RQ-1A Predator-A, Italian Air Force 32o Stormo, 28o Gruppo Velivoli Teleguidati, Foggia-Amendola Air Base, 2004

The Italian Air Force decided to acquire a squadron of Predators in 2000 based on the lessons from the recent peacekeeping operations in the former Yugoslavia. These became operational in 2004 at Foggia-Amendola Air Force Base. They are in the usual light ghost gray (FS 36375) finish with a tactical number on the fuselage side indicating the squadron and aircraft number, followed by the serial number and national insignia.

2: MQ-1B Predator-A, USAF 46th Expeditionary Reconnaissance Squadron, 333rd Air Expeditionary Wing, Balad Air Force Base, Iraq, 2004

The 46th ERS was deployed to Iraq in 2004 to provide "kinetic" capabilities to local field commanders, meaning hunter-killer missions rather than only reconnaissance. Their markings when first deployed were very austere compared to other Predator squadrons such as the 11th and 15th Reconnaissance Squadrons of the 57th Wing, which tend to be seen with the "WA" (Nellis AFB) or the later "CH" (Creech AFB) base codes. In this case, the markings are confined to the tail numbers AF/01 078, which indicate a USAF aircraft procured in Fiscal Year 2001, with 078 being the last three digits of its serial number. This serial number indicates that this Predator was originally built in the RQ-1A configuration, later rebuilt with the MQ-1B attack capabilities. The finish is typical of Predators; overall light ghost gray (FS 36375). The AGM-114 Hellfire missile under the wing pylon is in overall black with yellow bands indicating a war load with a high-explosive warhead.

E RQ-1 PREDATOR ENDURANCE UAV

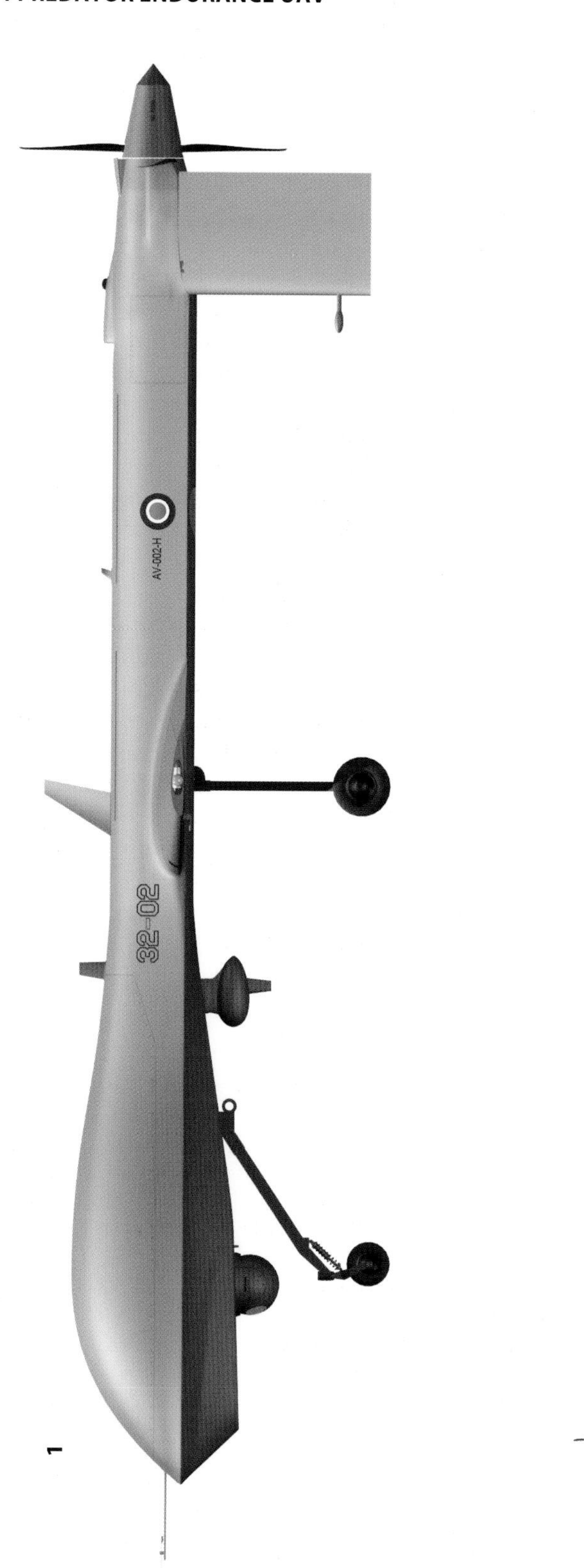

1

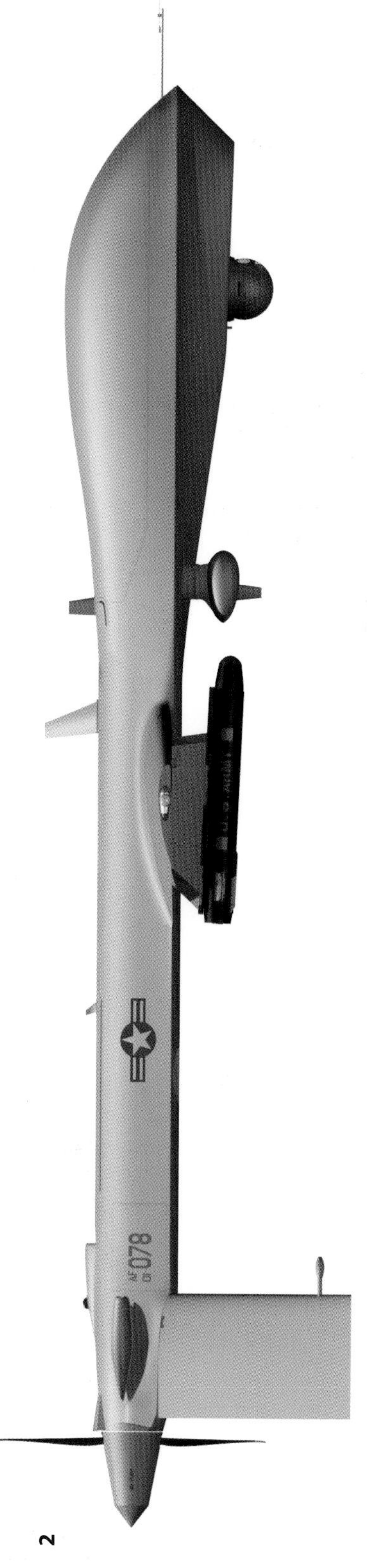

2

flight control system could be programmed to return to base even in the case of interruption. These new technologies at first were too large to be fitted on small tactical UAVs, but they made possible a new generation of long-range "endurance" UAVs in the 1990s.

The USAF was interested in long-endurance UAVs as an alternative to manned spy planes such as the U-2 and SR-71. In the 1970s, the Soviet Union began deploying the S-200 (SA-5 Gammon) antiaircraft missile, which was the first system capable of reaching the high-flying SR-71. While satellites had become the preferred method of strategic reconnaissance, they had several distinct limits. Their image quality was invariably affected by atmospheric obstructions such as clouds and humidity, and their flight paths were difficult and time consuming to alter. The first USAF strategic reconnaissance drones, like the ill-fated D-21 Tagboard, used autonomous guidance. With the advent of real-time cameras, there was more interest in command-guided endurance UAVs. The first steps in this direction took place in the mid-1970s with experimental work on the Boeing YQM-94 Compass Cope. This air vehicle used a sailplane configuration with a 90ft wingspan. The Compass Cope was large enough that it could carry a satellite uplink antenna. Although the Compass Cope did not proceed into production, it paved the groundwork for endurance UAVs two decades later. In the 1980s, the US Defense Advanced Projects Research Agency (DARPA) sponsored a number of endurance UAV demonstrations, including Boeing's Condor, Leading Systems' Amber IV, General Atomics Amber, and E-System's EVER. In 1988, the Department of Defense established a requirement for the UAV-E (UAV-Endurance) with a range of more than 1,000 miles and an endurance of about 48 hours for the targeting and surveillance of large areas of a battlefield or ocean.

Naval UAVs have been slow to arrive due to the difficulty of developing reliable methods to land the air vehicles automatically on deck. The US Navy is in the process of acquiring the RQ-8A Fire Scout for maritime applications. This example is seen during the first autonomous landing of a Fire Scout aboard ship, the USS *Nashville* LPD-13, on January 17, 2006. (US Navy)

The elegant RQ-4A Global Hawk is the largest and most sophisticated of the current robotic spy drones, capable of intercontinental flight and an endurance of over two days without refueling. (Northrop-Grumman)

In the short term, the DoD program had little effect since it typically took a decade between the first development contract awards and the actual fielding of an aircraft. When the Balkans crisis erupted in the early 1990s, the Pentagon saw the immediate need for an endurance UAV that could orbit over contested areas in the former Yugoslavia to monitor developments. Since it was unlikely that the cumbersome Pentagon procurement structure could react quickly enough, the program was handed over to the CIA to circumvent the bureaucratic delays. The CIA turned to General Atomics, which had bought out Leading Systems after the firm went bankrupt, acquiring their Amber endurance UAV in the process. This evolved into the Gnat-750, which had already flown in 1989. The Gnat-750 Tier 1 system was first deployed in early 1994 but lacked a satellite uplink antenna and so was used in conjunction with a Schweitzer RG-8 powered glider as the flying data relay substitute. In the meantime, the Pentagon continued to fund a more sophisticated version of the Gnat-750 with satellite uplink as the Tier-2 endurance UAV, later called the RQ-1 Predator. The Predator had a longer wingspan and a distinctive bulged nose that housed its satellite uplink antenna.

Miniaturized video cameras and microprocessors have made it possible to field "backpack" mini-UAVs for the infantry. This is an Aerovironment RQ-11 Raven being launched by a soldier from the 2nd Battalion, 27th Infantry Regiment during a search operation in Patika province in Iraq on November 16, 2006. (US Army)

The first Gnat-750 systems were deployed in 1995 in Albania to monitor the civil war in the former Yugoslavia. The USAF activated its first UAV unit since the Vietnam War on July 29, 1995, at Nellis Air Force Base, designated the 11th Reconnaissance Squadron of the 57th Wing. This unit formerly operated RF-4C reconnaissance aircraft and later served as a target drone control squadron with DC-130 aircraft. The first USAF Predator operations over the Balkans took place in February 1996 out of Hungary. During the first sortie, the Predator lost its satellite uplink signal but managed to fly back to its operating base at Taszar, Hungary, a testament to the value of the new GPS system. By 1997, the squadron was flying Predators on an almost daily basis, although the accident rate was very high due to the novelty of the equipment.

GROWTH AND EXPANSION

Besides the Predator, the USAF and CIA had teamed together to develop a stealth UAV that could operate in airspace threatened by air defense missiles. This system was dubbed the Tier 3 but did not progress very far before being cancelled due to concerns over its enormous potential cost. Instead, in 1994 the program was split in two: a very long-endurance UAV with a highly capable sensor payload called Tier 2+ and a smaller stealth UAV called Tier 3-. The Tier 3- contract went to the Lockheed Martin Skunk Works and resulted in the RQ-3 Dark Star, a flying wing design that looked like an unmanned B-2 bomber. Test flights began in March 1996, but the first prototype was lost on its second test flight. The program was cancelled in February 1999 due to cost concerns, although the later prototypes were flown for further demonstrations. The Tier 2+ contract was won by Teledyne-Ryan, later absorbed into Northrop-Grumman during the great aerospace industry consolidation of the 1990s. This design eventually emerged as the RQ-4 Global Hawk. Like the Predator, the Global Hawk used a large sailplane configuration. However, the air vehicle was jet powered for higher cruise

The USAF experimented with the armed versions of the Firebee as the BGM-34 in the early 1970s. This example is on display at the USAF Armament Museum at Eglin AFB and is seen armed with a free fall bomb. (Author)

speed and had a one-ton payload – three times greater than Predator's. Global Hawk first flew in February 1998, and the first series production aircraft was delivered to the 9th Reconnaissance Wing at Beale AFB in September 2003.

In the wake of the September 11, 2001, Al Qaeda attack in the United States, USAF and CIA UAV operations stepped up considerably in intensity. The initial combat operations in Afghanistan were preceded by extensive reconnaissance operations using USAF and CIA Predators, as well as by preproduction Global Hawks. Besides the reconnaissance versions, the USAF had been working on an armed "hunter-killer" version of Predator since 2000. By mounting a laser designator in the nose, the Predator could guide Hellfire missiles for pinpoint attacks. The first combat use of this feature took place in November 2002 when a Predator controlled by a CIA/USAF team in Djibouti destroyed a car carrying Qaed Salim Sinan in a remote desert in Yemen using a single Hellfire missile. Salim Sinan was the Al Qaeda chief in Yemen and had been responsible for the attack on the USS *Cole*. Some thought was given to arming the Global Hawk as well, but given its high-

The Boeing X-45A UCAV marked a revival in interest in assault drones after decades of neglect. This is one of its early demonstrations of its attack capability, dropping a GPS-guided bomb. (Boeing)

F RQ-4A BLOCK 20 GLOBAL HAWK, USAF 9TH RECONNAISSANCE SQUADRON, 2004

The original production batches of Global Hawks were finished in overall white. The subsequent production series has been finished in an overall gunship gray fuselage and white wings with markings in subdued black (FS 37038). As is the case with other USAF aircraft, the markings include a two-letter base code for the aircraft home base, which in the case of the 9th Reconnaissance Squadron is "BB" for "Beale AFB" in California, part of Air Combat Command.

TECHNICAL DATA

Length	44.4ft (13.5m)	**On-station endurance**	24 hours at 1,300 miles
Wingspan	116.2ft (35.4m)	**Loiter speed**	343 knots
Height	15.2ft (4.6m)	**Maximum altitude**	>65,000ft (>19,000m)
Takeoff weight	25,600lb (11,610kg)	**Propulsion**	Rolls-Royce AE3007H turbofan
Payload	2,000lb (910kg)		
Ferry range	15,545mi (25,015km)	**Sensor**	Raytheon Integrated Sensor Suite (ISS)
Maximum endurance	36 hours		

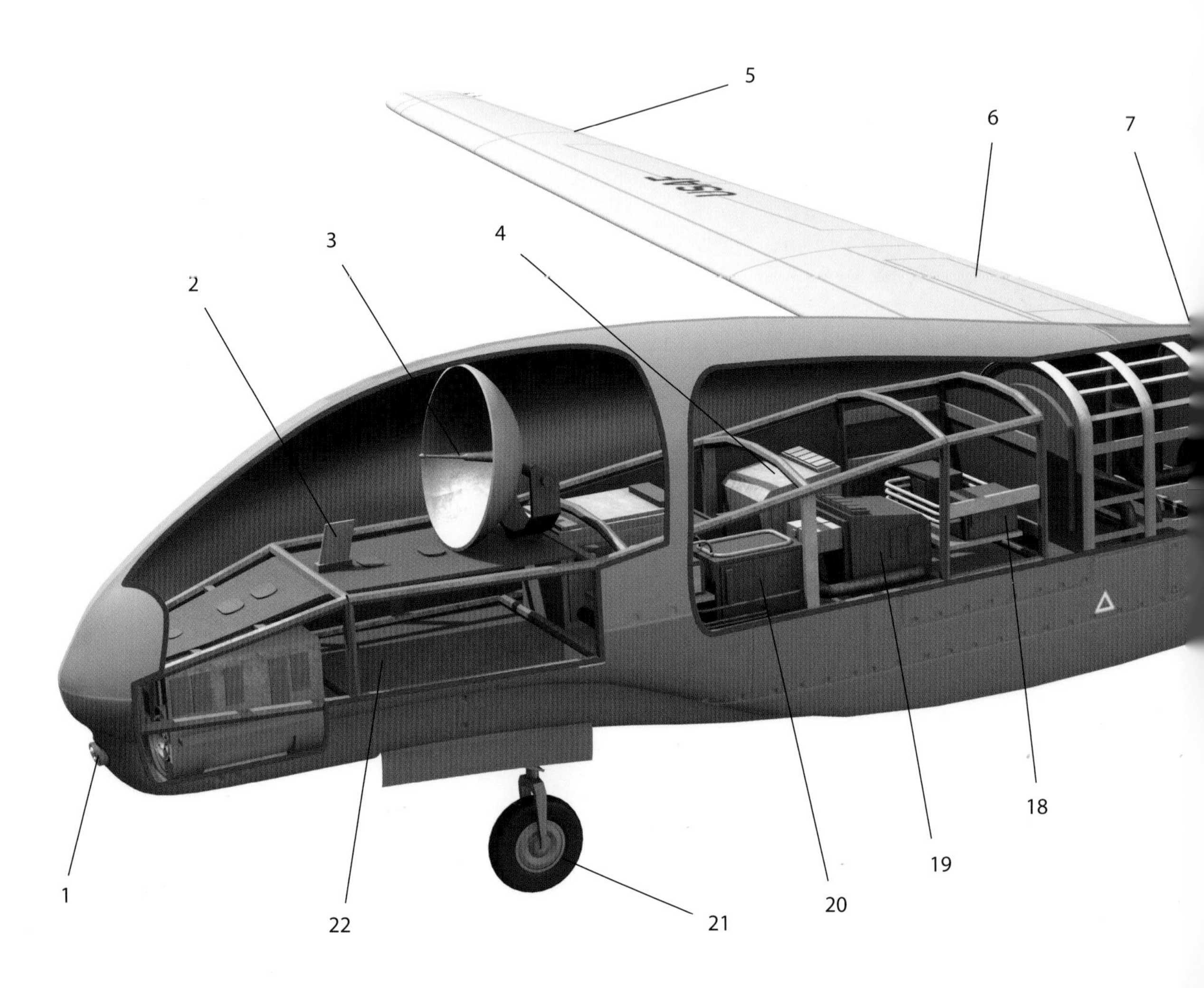

KEY

1	See-and-Detect collision avoidance camera port	**12**	Common datalink antenna fairing
2	UHF line-of-sight antenna	**13**	Undercarriage composite fairing
3	Ku-band satellite communication antenna	**14**	Outboard spoiler flap
4	Omnistar GPS satellite navigation system	**15**	Undercarriage door
5	Aileron	**16**	Port undercarriage
6	Spoiler flap	**17**	Undercarriage actuator
7	Main fuel cell	**18**	Solid state data recorder
8	Rolls-Royce A3007H turbofan engine	**19**	Common airborne moden assembly
9	UHF satellite communication antenna	**20**	EISS (Enhanced Integrated Sensor Suite) transmitter
10	Pitot tube	**21**	Forward landing gear
11	Engine air start ground connector	**22**	Conformal antenna bay

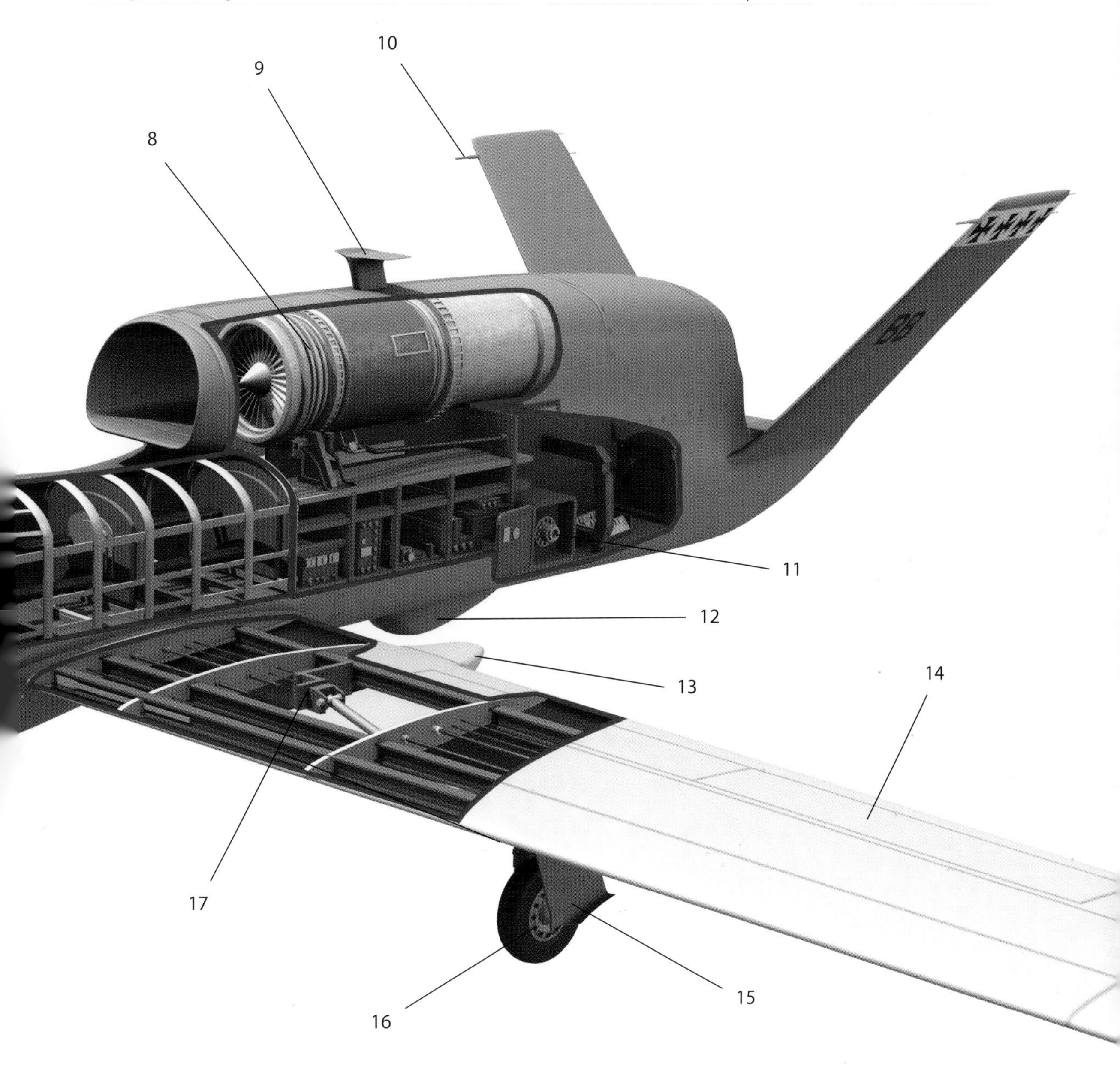

Helicopter assault drones are also likely in the future. This photograph shows a US Navy MQ-8 Fire Scout on trials firing Mk 66 2.75-in. rockets at the Yuma Proving Ground in July 2005. (US Navy)

altitude flight regime and certain treaty commitments, the USAF officially decided against an armed version. The utility of the hunter-killer Predator led to a later conversion program with all RQ-1 Predators being modified into the MQ-1 configuration with hardpoints for weapons as well as the necessary laser designator for weapons guidance. In addition, an upgraded and enlarged hunter-killer version was also developed, which emerged in 2006 as the MQ-9 Reaper. Since 2002, a variety of weapons have been tested for use on UAVs, including a version of the Bat smart bomb called Viper Strike and a variety of aircraft guided weapons including the Paveway laser-guided bombs.

The ensuing war in Iraq saw extensive use of many types of UAVs besides the endurance UAVs. After a decade of bureaucratic fumbles, the US Army had finally adopted a new tactical UAV, the RQ-7 Shadow. It was first used by the 104th Military Intelligence Battalion of the 4th Infantry Division during June operations near Tikrit. Over the next year, it was gradually

The first dedicated hunter-killer UAV is the MQ-9 Reaper, a derivative of the Predator. This is the first Reaper being delivered to the 42nd Attack Squadron at Creech Air Force Base, Nevada, on March 13, 2007. (USAF)

fielded with a number of other US Army units. The RQ-2 Pioneer was still in Marine Corps service, and it was deployed to Iraq as well.

The prolonged peacekeeping operations in Iraq and Afghanistan also saw the first extensive combat use of a new breed of robotic scouts, the mini-UAV, sometimes called back-pack UAVs since they are so easily portable. These are very small UAVs about the size of a model airplane. The US Army and US Marine Corps had first experimented with mini-UAVs in the 1980s with types such as the BQM-147 Exdrone and FQM-151 Pointer. These were used in small numbers during Operation *Desert Storm* in 1991. However, they were not especially effective as the early black-and-white cameras could not distinguish detail well enough in desert conditions, and their short range was a limit considering the long line-of-sight available in the western Iraqi desert where most of the heavy fighting took place. In the 1990s, these drones were continually improved with the addition of newer color video cameras and more compact ground control stations. Prior to the war in Iraq, the US Army had funded the development of a new generation mini-UAV, the RQ-11 Raven, for use by Special Operations Command. Likewise, the US Marine Corps had begun acquiring the similar Dragon Eye mini-UAV. These UAVs ushered in a new style of UAV operation since they could be deployed at small unit level to conduct routine patrols. They proved extremely handy to spot potential ambushes and also to search for the deadly IEDs (improvised explosive devices) favored by the Iraqi insurgents. The USAF used similar mini-UAVs for airbase perimeter search. Other armies taking part in the Iraqi occupation also used UAVs. Britain had problems operating its Phoenix UAV in hot weather conditions but used Predators and Ravens on loan from the United States, as well as Desert Hawk mini-UAVs. The NATO peacekeeping mission in Afghanistan also saw the extensive use of UAVs for patrols, including the use of the French SAGEM Sperwer by Canadian forces, Predator by Italian forces, and the Luna and Aladin mini-UAVs by the Germans.

Unmanned and unafraid. The first hunter-killer UAV was the MQ-1B Predator, a modified version of the RQ-1 Predator reconnaissance drone, adapted to carry the AGM-114 Hellfire laser-guided missile. It has been used in combat in Iraq, Afghanistan, Yemen, and other locations. (USAF)

The futuristic Northrop-Grumman X-47 Pegasus was the US Navy's entry in the Joint Uninhabited Combat Air System (J-UCAS) demonstration in 2004–06. The air intake design and delta configuration are stealth features designed to minimize its visibility to enemy radar. (Northrop-Grumman)

ASSAULT DRONES

The assault drone idea dates back to before World War II. There was very little interest in assault drones after the small-scale demonstration of the US Navy TDR-1 and TDN-1 assault drones in 1944. This was largely due to the focus on other robotic weapons, namely guided missiles. The idea of assault drones resurfaced in the final years of the Vietnam War when the USAF experimented with armed versions of the Firebee. Trials were conducted in 1972–73 using BGM-34 Firebees armed with a variety of guided weapons including the AGM-65 Maverick missile and Paveway laser-guided bombs. DARPA also experimented with armed versions of the Navy's QH-50 DASH helicopter drone. The Soviet Union conducted studies of an attack drone, the Sukhoi Korshun, which would have been controlled from an Su-24 mother ship had it been built.

The USAF and US Navy began to study assault drones, called UCAVs (uninhabited combat air vehicles), in the late 1990s. These were intended for use in environments too deadly for conventional aircraft, such as sites defended by advanced SAMs. Their principal mission is SEAD, although other missions are also under consideration. There was some hope that UCAVs could be built in large numbers at a cost significantly lower than combat aircraft since the airframe was smaller, there being no need for a cockpit. The USAF funded the Boeing X-45 while the US Navy funded the Northrop Grumman X-47 Pegasus. The program endured considerable turmoil, being combined into a joint program called J-UCAS (Joint

RQ-2 PIONEER TACTICAL UAV, US MARINE CORPS UAV SQUADRON TWO (VMU-2), IRAQ, 2003

When first in service with the US Marine Corps, the Pioneer was usually finished in a two-color gray camouflage scheme. However, aircraft in the field tended to become very beaten up, with numerous color patches, and in the mid-1990s, the Pioneers switched to a single color scheme in overall light ghost gray (FS 36375). Some Pioneers in Operation *Iraqi Freedom* in 2003 still had the older paint scheme, although often with panels and patches in light ghost gray. Markings are mostly in subdued black (FS 37038) except for warning markings such as the propeller warning on the tailboom. This air vehicle carries the squadron codes for VMU-2; VMU-1 carries "WG."

G RQ-2 PIONEER TACTICAL UAV

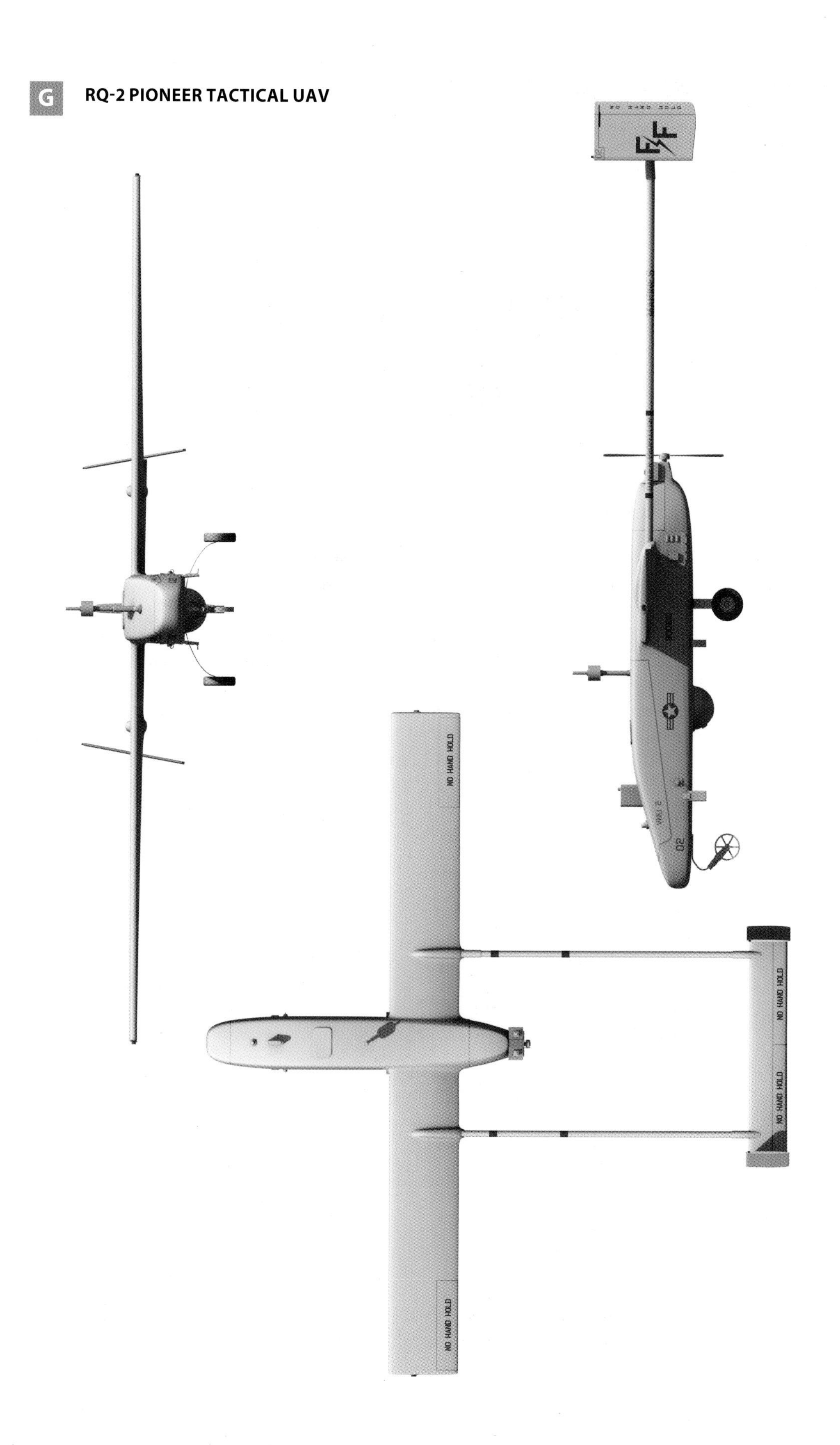

Russia's assault drone program has remained shrouded in mystery, although this Tupolev Tu-300 Korshun was displayed at the Moscow Air Show in 1995. It carries its weapons in an internal bomb bay or on an external pylon and has sensors on the nose for remote piloting and target detection. (Author)

Uninhabited Combat Air System) in 2003 and then being abruptly cancelled in December 2005. The Navy portion of the program continues with the Northrop-Grumman X-47B as a technology demonstration to examine UCAV operations from carriers, while the Air Force program is widely assumed to have gone "into the black" as a secret program.

There have been many suggestions that the current generation of combat aircraft will be the last generation with pilots onboard and that future generations will be remotely piloted. This is probably an exaggeration as UCAVs have not proven to be as cheap as predicted, and they seem poorly suited for some roles such as fighter aircraft. Nevertheless, UCAVs may very well begin to make inroads into combat aircraft production, and many countries have begun to fund development programs. Russia displayed a

France has been exploring hunter-killer UAVs with an armed version of its popular Sperwer tactical UAV, seen here in the armed Sperwer B configuration with missile launchers under the wing. (Author)

prototype UCAV, the Tu-300 Korshun (Kite), at the 1995 Moscow Air Show, and other programs are under development such as a UCAV version of the Yak-130 trainer called Proryv. France has been attempting to corral other European partners into a future European UCAV program called Neuron. Germany built a UCAV demonstrator called Barrakuda, but it crashed in 2006 early in its test program, and its future at the moment is uncertain. Italy has test flown a subscale UCAV demonstrator called Sky-X. Britain embarked its own Taranis UCAV in 2006 after having tested subscale demonstrators, and it is intended to replace the Tornado at some time in the future.

It seems likely that two types of armed UAVs will emerge – slow hunter-killers and fast, stealthy UCAVs. The hunter-killers such as the MQ-9 Reaper are primarily reconnaissance air vehicles that can attack targets when the opportunity arises. Flight control of these UAVs is by remote piloting from a GCS. The UCAV may be autonomous or remotely piloted. Lacking an elaborate sensor system, they may be preprogrammed to attack specific targets with little or no need for remote piloting. On the other hand, they may be designed with redundant remote-piloting capability to attack some types of targets such as moving targets or mobile SAMs that cannot be engaged using preprogrammed flight controls. Some of the more futuristic proposals have extended to hypersonic strike vehicles that could fly halfway around the globe, drop out of a flight path on the edge of space, and attack targets with precision munitions. The USAF considered robotic bombers for its Next Generation Long Range Strike (NGLRS) program, but at the time of writing, it will most likely be a piloted aircraft.

While UCAVs may represent the technological pinnacle of UAVs, considerable attention is also being focused on the smallest potential UAVs, the micro-UAVs. In the United States, DARPA has been funding the design of many flying vehicles that fit comfortably in the palm of the hand. Some mimic insect or bird flight while others resemble small helicopters. Many armies have expressed interest in these designs for scouting in urban environments, with the micro-UAVs being able to fly through buildings to scout for trouble.

France has been promoting a European UCAV program called Neuron. This is the mock-up of the air vehicle unveiled at the Paris Air Show in 2005. (Author)

Some significant technological hurdles remain before micro-UAVs prove practical. They are so small that it is difficult to fit them with transmitters and power sources with enough energy to transmit their video images through a building. Image quality at the moment is marginal due to the size and weight constraints. However, in time, it is possible that micro-UAVs will become as ubiquitous as other advanced infantry sensors such as night vision goggles.

THE FUTURE OF UAVS

On the modern battlefield, what can be seen can be destroyed. In an age of information warfare and network-centric battlefield tactics, something has to go out and gather the real-time intelligence. At the beginning of 20th century, real-time targeting was limited to the range of human eyesight. With the advent of UAVs, real-time targeting can be extended hundreds of miles into the depth of enemy territory. UAVs have proven to be the most versatile intelligence collection system of the current era, offering more flexibility and better image quality on the tactical battlefield than alternatives such as reconnaissance aircraft or satellites. UAVs are finally making possible the "reconnaissance-strike complexes" first dreamed about in the 1970s. They are a critical element in a battlefield network that begins by collecting real-

Britain is exploring a Strategic Unmanned Air Vehicle Experimental (SUAVE) as a potential replacement for the Tornado strike aircraft. In 2006, BAE Systems began development of the Taranis UCAV to further develop this technology. (BAE Systems)

The future of air warfare? Modern endurance UAVs are flown from ground control stations such as this Predator station of the USAF 46th Expeditionary Aerial Reconnaissance Squadron at Balad Air Base in Iraq in July 2004. (USAF)

time information on potential targets, then rapidly passes this information to a data fusion system that in turn rapidly processes the targeting information and passes it to the "shooter" best suited to attack the target, whether an artillery battery, tank, strike aircraft, UCAV, or missile system.

UAVs are still at the "bleeding edge of technology," perhaps most comparable to early combat aircraft at the time of World War I. Many configurations and missions have only recently been explored, and it will take at least a decade before many of these technologies are mature. At the moment, UAVs suffer from a much higher attrition rate than conventional aircraft, and until this issue is resolved by design improvements, the proliferation of UAVs will be limited. Another major technological and employment issue is the integration of UAVs into airspace traffic control since they pose a collision hazard with other aircraft. Sense-and-avoid technologies for UAVs have only recently begun to be studied. This challenge has an especially serious effect at slowing the proliferation of UAVs into civil aviation. In spite of these problems, UAVs are likely to be the fastest growing segment of the aerospace field over the next decade.

The advent of miniaturized computer processors permits mini-UAVs to be operated by laptop computers. This Marine from the 3rd Light Armored Reconnaissance Battalion is flying a Dragon Eye mini-UAV on a scouting mission from Camp Ripper, Kuwait, in March 2003 using special semitransparent eyeglasses that display moving map imagery of the mission while other imagery is provided by the laptop screen on his knees. (US Marine Corps)

New roles encourage new configurations. This Russian Splav R90 drone is designed to be fired from the tube of a 300mm Smerch multiple rocket launcher, after which it pops out its wings and scouts out targets for its host. (Author)

FURTHER READING

UAVs have attracted mountains of press coverage over the past decade, and there are several dedicated journals such as *Unmanned Vehicles* from the Shephard Group in the United Kingdom and *Unmanned Systems* from the Association for Unmanned Vehicle Systems International (AUVSI) in the United States. There have been few histories published about UAVs. Laurence "Nuke" Newcome's short paperback is the only historical survey, while Kenneth Munson's classic *World Unmanned Aircraft* is an encyclopedic treatment rather than a history. The two Wagner books cited below provide an excellent account of the Ryan Firebees, and Russian programs have been the subject of no fewer than three good histories.

Clark, Richard, *Uninhabited Combat Aerial Vehicles*, Maxwell AFB, Alabama, US Air University Press (2000)

Ganin, S. M., et. al., *Bespilotnye letatelnye apparaty*, St. Petersburg, Nevskiy Bastion (1999)

Gordon, Yefim, *Soviet/Russian Unmanned Aerial Vehicles*, Surrey, UK, Midland (2005)

Matusevich, N., *Sovetskie bespilotnye samolety-razvedchiki: Pervogo pokoleniya*, Minsk, Kharvest (2002)

McDaid, Hugh, and David Oliver, *Smart Weapons: Top Secret History of Remote Controlled Airborne Weapons*, New York, Barnes & Noble (1997)

Munson, Kenneth, *World Unmanned Aircraft*, Surrey, UK, Jane's (1988)

RIGHT
Future UAVs may bear little resemblance to piloted aircraft. The Honeywell Micro-UAV is being developed for the US Army's Future Combat System as a scout that can be flown from tanks and other mechanized vehicles. Unlike conventional UAVs, it can hover and stare at targets in an urban environment. (US Army)

Tactical UAVs can be recovered in a variety of ways, with parachute recovery being one of the most traditional approaches. To protect the UAV's electronics, an air bag system is often used to cushion the landing, as is seen here during the recovery of a Brevel air vehicle, the forerunner of the Bundeswehr's current Tucan tactical UAV. (MBDA)

Munson, Kenneth, *Unmanned Aerial Vehicles and Targets*, Surrey, UK, Jane's (various editions, 1995–2007)

Newcome, Laurence, *Unmanned Aviation: A Brief History of UAVs*, Reston, VA, AIAA (2004)

Wagner, William, *Lightning Bugs and Other Reconnaissance Drones*, Fallbrook, CA, Aero (1982)

Wagner, William, and W. Sloan, *Fireflies and Other UAVs*, Arlington, TX, Aerofax (1993)

Yenne, Bill, *Attack of the Drones: A History of Unmanned Aerial Combat*, Osceola, WI, Zenith (2004)

Zaloga, Steven, *World Unmanned Aerial Vehicle Systems*, Fairfax, VA, Teal Group (various editions, 2002–2007)

The current Russian Army tactical UAV system is the Sterkh system with its associated Yakovlev Yak-61 Shmel air vehicle that was accepted into service in 1997. The drone is launched from a version of the BTR-D airborne assault vehicle that also carried its launch rail on radio command guidance antenna. (Author)

INDEX

Figures in **bold** refer to illustrations.